The Atlantic Telegraph:

A

DISCOURSE

DELIVERED IN THE FIRST CHURCH,

AUGUST 8, 1858.

BY EZRA S. GANNETT.

Published by Request.

BOSTON:
CROSBY, NICHOLS, AND COMPANY,
117, WASHINGTON STREET.
1858.

The Atlantic Telegraph:

A

DISCOURSE

DELIVERED IN THE FIRST CHURCH,

AUGUST 8, 1858.

BY EZRA S. GANNETT.

Published by Request.

BOSTON:
CROSBY, NICHOLS, AND COMPANY,
117, WASHINGTON STREET.
1858.

The intelligence that the ATLANTIC TELEGRAPH CABLE had been successfully laid was received in Boston, on Thursday, Aug. 5, 1858. The following Discourse was delivered, on the next succeeding Sabbath, to the members of the Federal-street Church and the First Church, assembled in the First Church, Chauncy Street, — the two Societies having united, for public worship, during the temporary closing of the Federal-street Church. The Discourse is printed at the request of both Societies.

BOSTON:

PRINTED BY JOHN WILSON AND SON,

22, SCHOOL STREET.

THE ATLANTIC TELEGRAPH.

JOB xxxviii. 35: CANST THOU SEND LIGHTNINGS, THAT THEY MAY GO AND SAY UNTO THEE, HERE WE ARE?

THE surprise, hesitation, doubt, the belief, delight, enthusiasm, with which the intelligence that has been a principal subject of conversation the last two days, was received in our city, are a sufficient evidence of the interest taken in an enterprise, the practicability of which could be proved only by its success. That it has been attended with full success, is by no means clear; but so much has been accomplished, that ultimate failure may be pronounced impossible. Difficulties may remain to be overcome, before the submarine bond of union between two hemispheres can be used for the purposes of business and friendship, and unforeseen disaster may afterwards cause a suspension of the intercourse that shall have been established; but skill and patience must prevail at last. We are justified in considering the question virtually settled, whether the greatest discovery of this age can be put to the test of practical advantage on the grandest scale.

And a grand result it is, to be realized or anticipated, that the thought of man can travel through three thousand miles of space, with the rapidity of the lightning, and human hearts can use the bed of the ocean for the swift exchange of their sympathies. It cannot be a trivial event, at the annunciation of which worldly men broke forth into clamorous joy, and devout men bent their heads in grateful thanksgiving. It is not an improper subject to engage our minds on this day and in this place, since we regard it as the assurance of a series of benefits, whose extent and magnitude are beyond our power of computation. A new channel of blessing has been opened for the world. Is not this a suitable occasion for recognizing the providence and the goodness of God? Through a most unfortunate association of religion with the dark, rather than the bright aspects of life, any event which causes mourning is accepted as a fit theme for the pulpit, while the joyous side of experience is kept for the street and the domestic circle. What must be the consequence of invoking religion to add a deeper solemnity to whatever is sad, while our glad emotions are treated as unworthy to be brought into the house of worship, but to implant false views of the providence of God, and to restrict the domain of faith? Why should death be more sacred than life? Why should the accident, as we term it, — the result probably of human negligence, — by which a thrill of horror is sent through the land, be made the occasion of address, appeal or warning in numberless discourses, but the achievement by which the community is thrown into a transport of delight, — the fruit of a wise use of the Divine laws, — be held as too secular for the purposes of Christian instruction? Away with this ungenial superstition! God made us to be glad, and we do not profane his temple when we bring our gladness within its walls. "Oh, that men would praise the Lord for his goodness and for his wonderful works to the children of men. Let them exalt him also in the congregation of the people, and praise him in the assembly of the elders."

The completion of the work in regard to which there has been so much room for incredulity, is suited to awaken our religious feelings, not only through its connection with the Providence that includes all efforts and all results, but in view of the illustration it affords of the mysterious yet beneficent method which that Providence observes in its care of the world. Wonderful as are the agencies of which the Atlantic Telegraph may be taken as the consummation, they are only the product of human skill availing itself of means which the Almighty had prepared for its use from the beginning of the creation. The science which has laid hold on one of the most subtile elements in nature, and bound it in perpetual service; the art which has constructed the bonds of its captivity, the materials which have been woven to be at once its chain and its pathway; the depths of the sea, which for the first time in the history of our race have been made subservient to the uses of humanity; the electric fluid with which continents will hold communication through future ages; and the mind of man, which has called science and art to the task of combining the palpable, the invisible, and the intangible, into a means of helping the world to reach its goal, — all these, and that final destiny to which the world is brought so much nearer by this means, existed from the first in the prolific thought of God. Delighted and astonished as we are, we see nothing here which the

Eternal Mind did not foresee and arrange when He laid the foundations of the earth. The labor and ingenuity of man have accomplished no more than was intended. And it has been accomplished at the intended moment. Of all and each of the great agencies for human good, who can doubt, that they come "in the fulness of time"? By all his inventions man adds only new forms to the universe. Substance and ability are from the Uncreated One. The fabrics which man fashions into the garments of his thought, are furnished by Him who made all things. Nor do any of his inventions anticipate the Divine wisdom. From the simple raft to the steam-ship with its complete equipment, from the alphabet to the printing-press, from the lowest point of savage ignorance to the highest reach of scientific inquiry, every stage and step of improvement has but helped to fulfil and unfold the design of the world's Author. Beautiful and instructive is this evolution of the eternal scheme, through the energies which now want, and now opportunity, excites to action. The connection of the Old World with the New in the middle of the nineteenth century is as strictly providential, as was the discovery of America at the close of the fifteenth, or the arrival of a body of Christian emigrants on these shores near the beginning of the seventeenth. Marvellous and gracious is the providence beneath which the generations work out their unforced, yet predetermined history. The properties of matter and the laws of the physical universe were the same before Morse, Franklin, or Newton studied them, before chemistry had a pupil, or astronomy a name, that they are now. Nothing is new with God; nothing original with man. When the intelligence came for which we had been waiting with more of despair than of hope, our faith in God received confirmation, and piety felt itself strengthened for its conquest over wickedness.

To God be the glory! And yet man is exalted by the success which has crowned his undertaking. How forcibly are we reminded of the command given to him who was created "in the image of God," that he should "subdue" all sublunary things to his own will. What an example of man's power over nature does this achievement present! Space and force offer no hindrance to the execution of his wish. Neither the breadth nor the depth of the ocean is an insuperable barrier to the realization of his idea. The impracticable becomes unknown to effort, the impossible disappears from the horizon of hope. It was but as yesterday, when the question of four thousand years ago might have been repeated in scornful rebuke of man's presumption, — "Canst thou send lightnings, that they may go and say unto thee, Here we are!" That question can now be answered with an affirmation so literal that the accuracy of its terms almost startles us. Yes; the invisible, imponderable substance, force, whatever it be — we do not even certainly know what it is which we are dealing with, — the swift-winged messenger of destruction, the vital energy of the material creation, is brought under our control, to do our errands, like any menial, nay, like a very slave. Who shall now describe the circle within which human ability must confine itself? In neither height nor depth, in neither the seen nor the unseen, in neither the forms nor the elements which compose the universe, neither in that which terrifies the barbarian nor in that which perplexes the philosopher, do we recognize a limit to our subjugation of nature. What an amount of proof has the last quarter of a century furnished, that man was made to rule the earth! Not by force, nor by cunning; but by honestly and patiently gained acquaintance with the constitution of the world in which he lives. Science is the arm which he stretches out, and the hand with which he seizes on the mysteries that lie around him. The conqueror now is not he who leads armies to the battle-field, but he who by calculation and experiment widens the domain of knowledge. The student and the engineer, in our age, win a success that shames the victories of Napoleon. What can stand before the determination of modern science? What can baffle its keen-eyed and persistent scrutiny? Nor is it unworthy of notice, that the spoils which science now brings home from the fields on which it clothes itself with glory are laid at the feet of utility. The day of speculative inquiry is past. The acquisitions of the scholars who penetrate the recesses of nature, where they find richer treasures than the cities of India or the streams of California yield to their invaders, are freely offered to the world's use. Mind asserts its superiority over matter, not in a spirit of self-admiration, but for the sake of enriching life. Our scholars are the true philanthropists of the age. We have an example of the justice of this remark, in the enterprise which has determined the present direction of our thoughts. Its success is due to a union of the most acute scientific investigation, with careful and ingenious mechanical labor; both drawn to the work, not by an empty ambition nor by the hope of pecuniary reward, but by a desire to serve the practical interests of the two countries immediately concerned — the people, let us observe, and not simply nor chiefly the governments of those countries, — and through them of mankind. It is a noble tribute to the utilitarian spirit of the age, the spirit which requires knowledge to justify itself by promoting the general good.

It is not without its significance, that the scene of the two grandest exhibitions of the benefit which may result from a combination of scientific principle with mechanic art is the ocean, that part of the earth, which, less than the solid land, less even than the atmosphere, might seem to be within the domain of man. The most perfect display of his inventive and practical genius is a ship, bearing in its form the massive strength with the graceful proportions which attest the right of naval architecture to be considered one of the fine arts, while in the forest of beauty over which the eye wanders, from the solid trunk of the mast to the feathery lightness of the topmast rigging, it rivals the sculptured elegance of the Corinthian temple or the Gothic cathedral. Behold this marvel of man's production, whose weight, whose dimensions, whose capacity seem to entitle it to be called the monster of the deep, which yet sits with such a queenlike dignity upon the waves that we imagine them to be proud of their burthen; the home of a thousand men, who find comfort and luxury in its internal arrangements; the witness of a city's wealth, the champion of a nation's honor, a proof of the world's advancement; see this almost living creation of man's brain and hand, careering over the Atlantic with an air of conscious sovereignty. It is a spectacle that may well fill the heart of man with admiration of the place which he holds among the works of God; the ocean his highway, and the winds his ministers. But now from the ocean's surface, sprinkled all over with the evidences of its submission to man's purposes, look down through miles of liquid distance to its bed, and there see the last result of human toil, stretching its immense length across that bed, which never before since the earth was made had yielded a single foot to man's use, but now is doomed to own his dominion through the centuries of the world's future history, as the loves and fears of human hearts shall rush over its subject spaces. Tell me, my friends, where you will find more signal proof of man's greatness, than in these two examples of the ocean's subserviency to his convenience.

Still it would be a small recompense of the time and labor and anxiety that have been expended in perfecting this last expression of man's inventive thought, if it only supplied fuel to his self-love, or even gave him a higher estimation of his own capacities. The temper of the times, I have said, is essentially practical, as the history of this enterprise may show. Though necessarily conducted by scientific men, it originated in a conviction of the commercial and political advantages it would yield, and was sustained through its expensive period of preparation by those in private life and in public office, who were looking at its effects on social and national interests. The enthusiasm with which the news of its completion was received, arose from a deeper source than the joy we always feel when anxiety is displaced by an assurance of success. The mind at once caught a glimpse of its distant as well as its immediate consequences, — its connection with social progress, the world's civilization, and the unity of mankind. And now, when we lay our sacrifice of praise on the altar of God, we think of the benefits which will flow from this triumph over difficulty and disappointment. To enumerate those benefits would be an office which prophecy alone could undertake. The good contained in the future is wisely concealed from us, as well as the evil. No one can tell what changes in the relations of the two portions of the civilized world to one another, and through them in the relations of all the countries of the earth, may follow upon the establishment of instantaneous communication between the two continents. We may draw a lesson from the fact, that remarkable inventions or discoveries have generally produced effects very different from those which were anticipated, when they first came into use; what then seemed to be merely incidental or insignificant becoming the most valuable, while the results which were fondly contemplated have sunk into subordinate importance. It is not, therefore, an indication of wisdom, to speculate with much confidence, on the fruits of any enterprise at its commencement. The innocent author of the slave-trade had as little foresight of the evils which he was introducing, as the constructer of the first steam-vessel of the benefits which he was conferring on the world. Some conception, however, may be formed of the bearings of telegraphic intercourse between the Eastern and Western Continents on the interests of humanity. Its final influence on mercantile affairs as no one can predict, so its immediate effect only they who are familiar with the details of trade can describe. It is plain that the transmission of intelligence with a rapidity that must put an end to crafty calculation on the one hand, and to anxious suspense on the other, will tend to give a more healthy tone to business. When every day shall renew the solution of the paradox, that the merchant in Boston or New York may learn as he enters his counting-room in the morning, what were the transactions of the Exchange in London or Paris at noon, a check will be placed on that wild spirit of speculation, by which the comfort and the morals of life have been so much impaired in this country. On a higher plane of advantage we may mark the relief in numberless cases, the satisfaction every moment arising from the knowledge friends may obtain of one another's condition almost without delay, although separated by thousands of miles. Imagine your child or your parent in Europe, and, instead of a letter's tedious travel across the Atlantic, a message flying through the electric wires; will not space and time be virtually abolished? What a change, when the tenant of a house in England or Germany shall be my next-door neighbor, and soon Constantinople and Calcutta shall be but a street's length from my own dwelling! On political affairs the influence of this means of international conversation cannot but be important. How many misunderstandings may be prevented from growing into serious disputes, how many jealousies be dispelled, by a few taps of a little machine; how much expense be saved by the use of an invisible, yet faithful messenger, in place of a special diplomatic agent who might only succeed in "darkening counsel by words." It is one of the happiest auspices of the future, which it belongs to our age, and may we not say to our country, to have introduced, that the intercourse of governments has been taken out of the mystery of equivocation in which the practice of ages had wrapt it, and been in great measure reduced to plain and frank speech. The telegraph will compel statesmen to study yet greater simplicity of expression.

The connection of Great Britain with our own shores is not a fact that can stand alone. By this channel we shall hold communication with Continental Europe, Asia, Africa and Australia. The line of electric transmission will soon girdle the globe. Civilization must receive an impulse greater than it has felt at any time within the last half-century, memorable as this period has been for the appearance of agencies suited to awaken intellectual activity. How can we bring before our minds, with sufficient distinctness, the consequence of a universal interchange of thought by the speediest method? The world, it has been said, will be made a great whispering-gallery; I would rather say, a great assembly, where every one will see and hear every one else. The press has for the last fifty years been the chief agent in educating society. It must now share that privilege with another instrument. The telegraph will anticipate the journal. Facts and opinions, the materials with which successive generations construct the road along which society advances with unequal, but never retrograde steps, will now be furnished to every one on the moment. What an excitement will be given to the brain and heart of the world! Too much, you may say. Perhaps so. Excess, however, will not be fatal. Society will accommodate itself to the new conditions of its existence; and then improvement will go on steadily, as well as rapidly.

The most remarkable effect, if I may judge from my own narrow thought, will be the approach to a practical unity of the human race; of which we have never yet had a foreshadowing, except in the gospel of Christ. Actually, the race has been divided into as distinct portions as if they lived on separate planets. Jealous of one another, or mutually unknown, they have exchanged no sympathies, united in no common labors, recognized no obligations of kindred blood. What has China been to the rest of mankind for hundreds of years? Even on the maps of the geographer, what has the interior of Africa been, though now known to contain populous cities, but an arid desert? Can such ignorance and isolation continue after the lightnings shall have been taken into the service of man, to go hither and thither at his command, saying, Here we are? The death blow has been struck to barbarism. An exclusive policy must yield to the universal solvent. The telegraph is cosmopolitan. Not more British than American, it can neither be monopolized by government, nor stopped in its work of civilization by neglect. It is an institution for the people. Its office is, to diffuse intelligence; its effect, to allay differences. Men who talk together daily cannot hate or disown one another.

Learning and freedom must be promoted by this disperser of darkness, this intruder into the haunts of senile conservatism. By many persons it is thought that an increase of religious faith and life must follow. But this effect seems to me more doubtful, because I do not perceive any direct bearing of an open intercourse among men on spiritual character, and because facts show that neither learning nor freedom is the sure precursor of virtue. We may indulge the hope, that men will not abuse an increase of light and opportunity, but their past history should teach us that such abuse is not impossible, nor improbable. God unquestionably meant, that all our powers and all the circumstances of our being should bear the seal of consecration: but men do not believe with a firmer faith because they have more light, nor do they render a more strict obedience because it is made to depend on their own will. One of our most common mistakes consists in imputing the end as a certain consequence of an enjoyment of the means; whereas knowledge is often a cause of increased vice and misery. The greater the ability, the greater may be the wickedness. The introduction of a new means of improvement and happiness should render us more watchful over ourselves, rather than less faithful. The most intelligent community is not always the most moral. The common school is not an inevitable blessing to all whom it gathers under its instruction. It may be an effect of augmented facilities of communication, that men will become less patient, less contented, less mindful of the Power that upholds their lives. It is not good to live too much abroad, too much in society. Retirement and repose are needful for the maturity of religious character, as both the little flowers and the great trees need night as well as day for their growth. There is danger, therefore, that the circumstances which might assist, will impede our excellence. Religion must have its root in the soul; it cannot flourish as a parasite on some other stock. Religion is spiritual experience, not mental development. A Christian is not one who cultivates his powers to the utmost, but one who subjects his whole nature to the guidance of Divine truth. Two different, but not very dissimilar forms of error captivate many persons in our day. One represents honesty of purpose and activity of life as the fundamental laws of our being, and repudiates faith as a weakness or a misfortune. A body who have taken to themselves the name of Secularists, that there may be no disguise over their purposes, have for some time been busy in propagating their opinions in England, through the efforts of a man of undeniable sincerity and ability who stands at their head. The want of a similar organization or name does not prevent the existence of the same class of minds in this country. The other error induces a reliance on education, taste and physical comfort for the production of a spiritual character; as if the buttresses and ornaments of a building could take the place of substantial walls. There are reformers among us, worthy to bear the title, who, adopting this error, fall into the delusion of confounding philanthropy with piety. Now it is clear, that men may be highly educated, and yet lack both conscientiousness and devotion; not thoroughly educated, of course, for the noblest part of their nature is neglected; but educated according to that false use of the term — so common, alas! here and everywhere — which confines it to mental accomplishment. It is equally clear, that a community may possess every means of outward prosperity, and yet be corrupt at heart. Let us, then, not be caught by the fallacy which substitutes material and social blessings for personal righteousness, issuing from the fountain which Christ opens within the believer, and which "springeth up into everlasting life."

After such cautions I need not be timid in the expression of a belief, that the ultimate, if not the immediate effect of that discovery, of which we are now gathering the first ripe fruit, will be a diffusion of Christian principles and sentiments through the earth. It is not self-conceit which moves us to notice the fact, that the two countries which are not only united by a chord along which every pulsation in the heart of one may be borne to the other, but which joined their efforts in laying that chord where neither the hand of violence, nor, as we may hope, the chafing of time can reach it, that these two countries, I say, stand in the front rank of Christian nations, — the representatives and guardians of the most vital forms of religion on earth. Is there not a meaning and a prophecy in this fact? May we not believe that the Providence which uses human freedom for the fulfilment of its own designs, and which announces the redemption of man from all evil as its final design, has committed to these two nations the sacred trust of bringing mankind into harmony of faith and life?

The signs of the times may appear to some eyes gloomy. But to the believer in God's eternal purpose of regeneration for the race which was "made subject to vanity by reason of Him who subjected the same in hope," the hour is auspicious, for it betokens union, progress and success. I do not ask you to indulge any unreasonable expectations. I do not say that the next intelligence from the island, which has sent its first message as a vibration of joy through our land, may not overwhelm us with disappointment; but not such as shall last. No. Enough has been done to be taken as a pledge that the work will neither be relinquished nor defeated. I dare not hope that the amicable relations which now exist between our Republic and the British Empire will never be disturbed by a feeling of jealousy or animosity. The time when wars shall cease, has not come. I am not blind to the symptoms of uneasiness and misunderstanding between the governments of Europe. But why need we look only on omens of disaster? Why construe outward courtesies as a mask to hide secret hatred? Why shall we distrust, not only the generous sentiments, but the sound policy by which rulers profess to be governed? Why not have a little more faith in man, and very much more faith in God?

My friends, let us hail the event, which we have thought it no profanation of the sanctuary to make the subject of our discourse, as the harbinger of a closer sympathy and a more effective co-operation by which not only England and the United States, but the nations of Christendom, and soon all the people of the earth, shall be bound together in the interchange, not of friendly messages alone, but of kind offices and fraternal regards. In one of our cities, we are told, the indubitable confirmation of this event will be announced by the sweet music of bells ringing out their anthem of praise. Let us forerun that chime of metallic voices, while with our tongues and hearts we celebrate the triumphs of science, the goodness of God, and the hopes of humanity. It is no more than sober anticipation, which sees in the future history of our race achievements of mind over matter beyond any that have yet been chronicled, blessings scattered in wider profusion than has ever yet gladdened the eyes of the philanthropist, and a social progress in comparison with which the movements of the present age shall seem slow. It is not a dream of the fancy, but a lesson of our faith, which we interpret into a promise of universal civilization and universal righteousness. The era of peace and goodwill, is it a vision never to be realized? The kingdom of Heaven embracing and sanctifying the various communities of mankind, is it a mere figure of speech? I trust in God it is not. The world will be redeemed from the bondage of corruption. The gospel of the Father's love will run through the earth, and win all souls to a grateful obedience. Hope, prophecy, faith, they are not blind teachers of the blind. It was in an hour of holy exultation that Jesus exclaimed, "I beheld Satan as lightning fall from heaven." The overthrow which to his vision, familiar with the measures of eternity, appeared as swift and as resplendent as the lightning's course, centuries of patience and struggle will yet be required to effect.

But it is as sure as the Eternal Word. May not we take the lightning as a type of its certainty, when we see that, which has been the world's admiration and dread, converted into the messenger of its pleasure and the minister of its profit ?

www.ingramcontent.com/pod-product-compliance
Lightning Source LLC
LaVergne TN
LVHW020636110826
845149LV00004B/1223

* 9 7 8 1 4 1 8 1 9 1 6 2 7 *